# NOTE-FINDER SYSTEM FOR DESCANT RECORDER

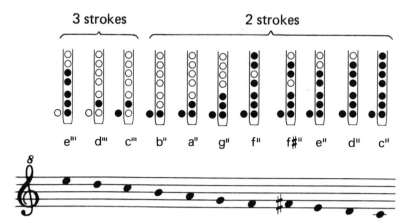

3 strokes     2 strokes

e'''   d'''   c'''   b''   a''   g''   f''   f#''   e''   d''   c''

Standard notation, 5 lines and 4 spaces. Each note has its own 'note-finder', showing which holes must be covered. Strokes ( '' or ''' ) after a note's name indicate the pitch.

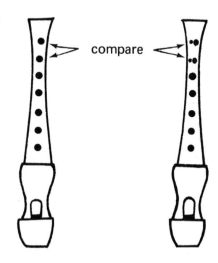

compare

*Some holes on your recorder may be double ones,* but this does not affect the fingering diagrams ('note-finders').

# NOTE VA

GW00580105

Here you *read* and *tap out* the *rhythms* si[...]
Figure or letter *not* in brackets = *read* and [...]
Figure or letter *in* brackets = read but *don't* tap.

| Tap : | o | | | | = SEMIBREVE (whole note) |
|---|---|---|---|---|---|
| Count: | one | (two) | (three) | (four) | |

| Tap : | ♩ | | ♩ | | = MINIMS (half notes) |
|---|---|---|---|---|---|
| Count: | one | (two) | three | (four) | |

| Tap : | ♩ | ♩ | ♩ | ♩ | = CROTCHETS (quarter notes) |
|---|---|---|---|---|---|
| Count: | one | two | three | four | |

| Tap : | ♫ | | ♫ | | ♪ ♪ | = QUAVERS (eighth notes) |
|---|---|---|---|---|---|
| Count: | one & | two & | three & | four & | (eighth notes) |

# RESTS

Rest signs indicate the same length in time as their equivalent note values (see above), so you can count them in exactly the same way.

| Semibreve rest (whole note) | Minim rest (half note) | Crotchet rest (quarter note) | Quaver rest (eighth note) |
|---|---|---|---|

# SOME ADVICE WHEN PLAYING YOUR RECORDER

Follow it and you'll find playing so much easier. *And* it'll sound a lot better!

## How to sit

Sit upright on a chair without arms. Sit up as straight as you can. That's right – arms loose from the shoulder, wrists straight and elbows held a little bit away from your body. All this, you see, makes it easier to breathe. And your fingers will move more easily, too!. When you play standing up, then stand in the same way.

## Here's how to cover the holes

Cover the holes with the centre of the pads of your fingertips. Cover properly, otherwise the air will get out and you won't get a clear note, or perhaps no note at all.

Curve your fingers slightly. Then place your right thumb opposite your middle finger.

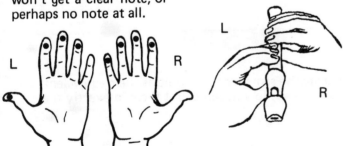

## Here's how you blow

Blow one note, softly and evenly. You may find some difficulty to begin with in blowing at the right pressure. Never mind. You'll soon get the trick of it!

## The different parts of your recorder

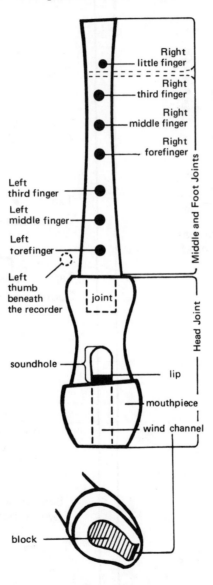

Right little finger
Right third finger
Right middle finger
Right forefinger

Middle and Foot Joints

Left third finger
Left middle finger
Left forefinger
Left thumb beneath the recorder

joint

Head Joint

soundhole
lip
mouthpiece
wind channel

block

**Descant recorder**

## Position of mouth

Pout a little, as if you were going to suck your thumb (surely you haven't forgotten how to do that?). Now place the mouthpiece in the cone formed by your lips. Be sure to leave a little space between your front teeth and the mouthpiece, otherwise your tongue won't be able to do its job properly.

N.B. Position of Right thumb

## This is how the tongue works

Always start and stop a note with the tip of your tongue, as if saying 'toot' (but don't 'spit' the 't' sound, or the mouthpiece will get blocked with spittle and prevent the sound coming out clearly). As you do this, the tip of your tongue must move between two positions:

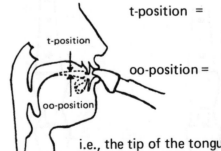

t-position
oo-position

t-position = tip of tongue is placed immediately behind the *upper* front teeth.

oo-position = tip of tongue is placed immediately behind the *lower* front teeth.

i.e., the tip of the tongue closes and opens the airstream. When you want to blow several notes *rapidly* one after another, you put them together like this: 'tootootoot.'

# Play a Tune

is a song and recorder book, with guitar accompaniments.

*IT'S MEANT FOR YOU* who've never blown a note on a recorder in your life, *and* for you, with a beginner's knowledge of the guitar.

*IT'S DIVIDED UP INTO TWO SECTIONS:*  An Accompaniment Section (pp. 2-22)
A Melody Section (pp. 23-30)

*RECORDER PLAYERS* start by playing the accompaniment lines in the Accompaniment Section. This will gradually provide them with the notes they need for the tunes in the Melody Section— (see the arrows below in 'Tunes in order of playing'). Also read 'Recorder tips' when they appear in the book and the 'General method' on p. 3.

*GUITARISTS* can play melodies and accomps of various types and grades of difficulty. Also read on p. 32 and inside the back cover. Also 'Guitar tips' when they appear in the book and 'General method' on p.3.

## Tunes in order of playing

So here are three of our commonest domestic musical instruments - voice, recorder, and guitar. And here's to you who'll be using them! TUNE UP!

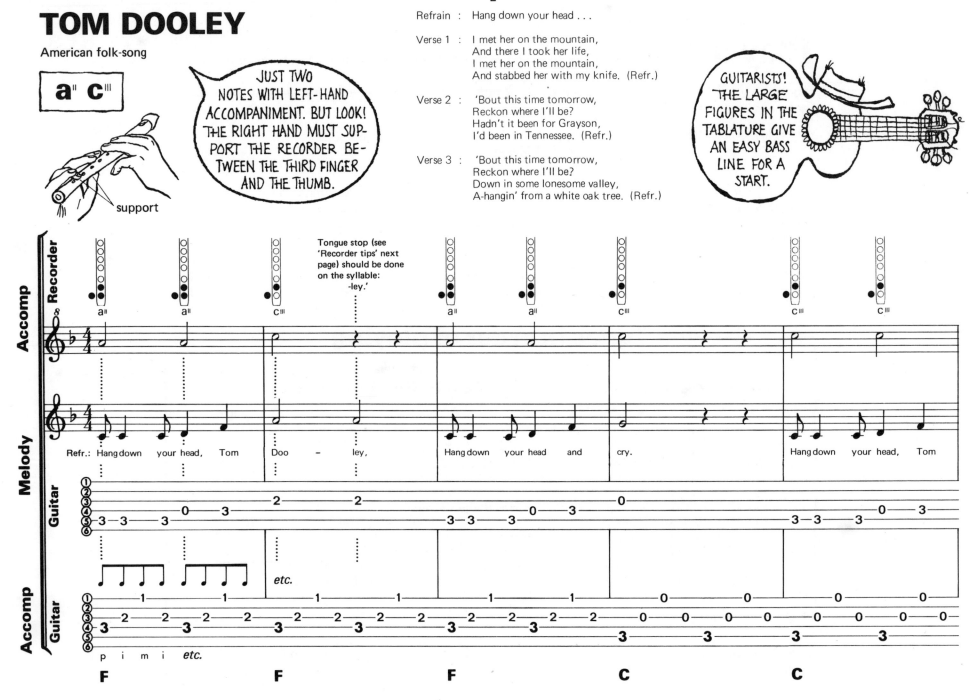

**Recorder tips**

1. Blow the notes quickly, one after another, leaving almost no gap at all between them.

2. Before each break stop the note with your tongue (like the 't' in 'toot') = *tongue stop*.

3. If you like you can also play the accomp with the same rhythm as the melody.

**Guitar tips**

Read p. 32 and inside back cover; also read 'General method' on this page.

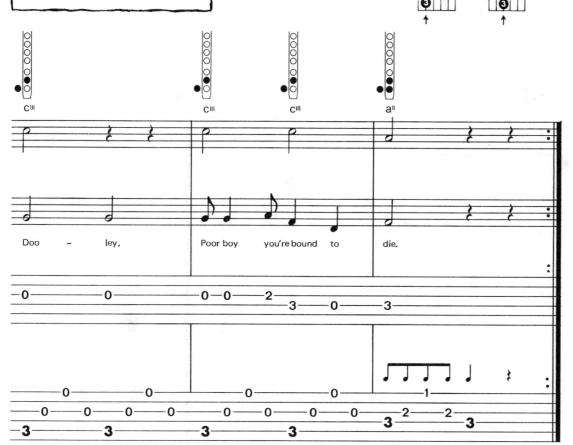

Doo - ley,   Poor boy   you're bound   to   die.

# General method

My last will and testament is that you shall read the following and remember it.

## Rhythm

Run your eye upwards from the bottom to the top of the whole system (see dotted lines on previous page). The visual image agrees in every respect with what is going on rhythmically, in tune, in words, and in accomp. What I mean is, you can *see* the length of each note just as you can *hear* it (you'll also find a counting method presented on p. 7).

## Recorder

Switch to and fro between the Accomp Section (pp. 2 – 22) and the Melody Section (pp. 23 – 30). See 'Tunes in order of playing', Preface, p. 1. Also read the 'Recorder tips' as they crop up in the tutor.

## Guitar

Read p. 32 and inside back cover. Also read the 'Guitar tips' as they crop up in the tutor.

Deep in the Wild West 3.2.75

In all haste, yours   *Tom Dooley*

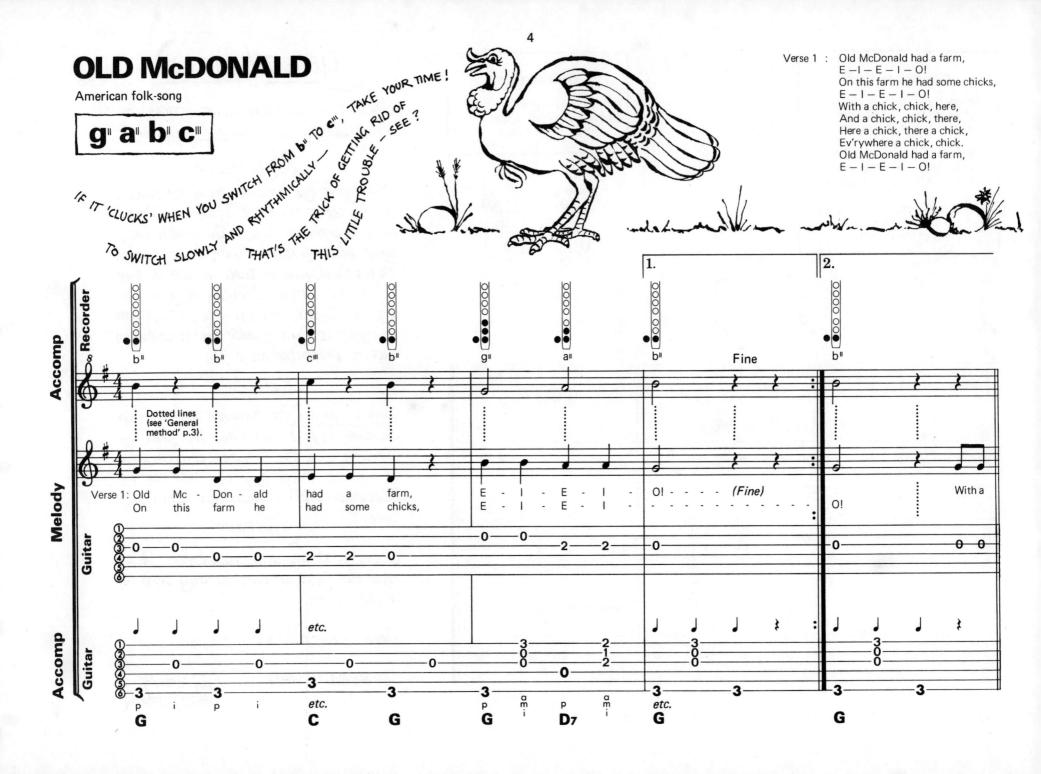

# OLD MAN SINGH

Melody and words by Lasse Sundberg
English version by Paul Britten Austin

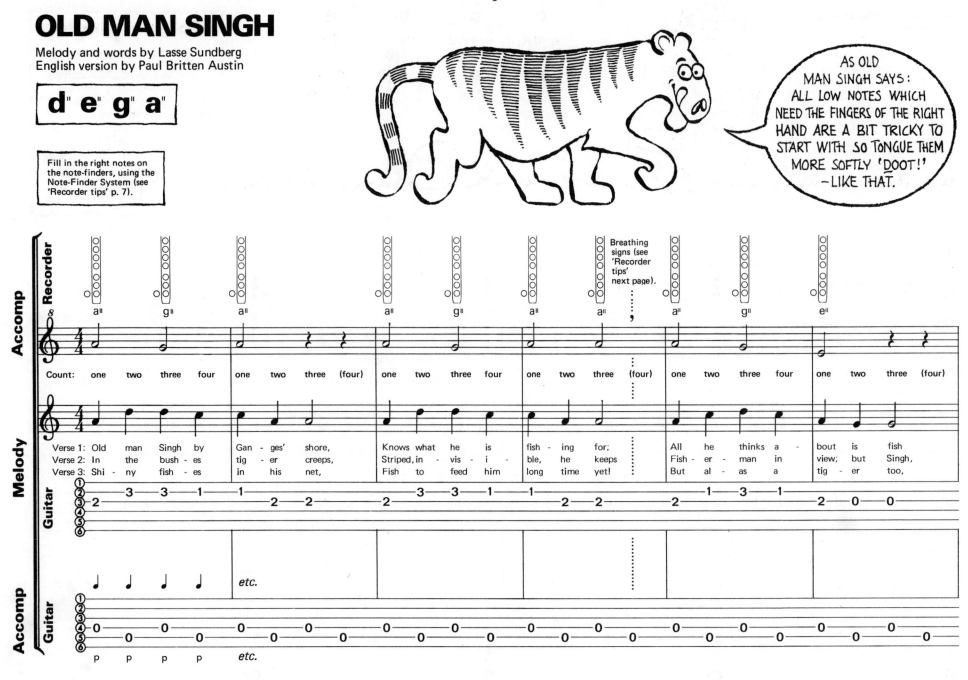

| | | | |
|---|---|---|---|
| d″ | e″ | g″ | a″ |

| one | two | three | four | one | two | three | (four) |
|---|---|---|---|---|---|---|---|

Does - n't hear the bush - es swish.
Does - n't no - tice an - y - thing.
Hates a di - et of bam - boo.

## Guitar tips

The whole character of this tune demands that you don't accompany it with traditional guitar chords. It just cries out for 'something different'. For instance:

1. *The tablature accomp,* with its obstinately repeated alternation of strings ④ and ⑤.

2. *The diagram chord* below, which you can go on playing throughout the whole piece and in any rhythm you like.

p〜〜 = the thumb is drawn across the strings in the direction of the arrow = *arpeggio.*

## Recorder tips

1. *Fill in the note-finder notes for the accomp line,* using the Note-finder System on the pull-out page. Pay attention to the position of the note's 'head' (whether it lies across a line or between two lines); also its name *(including whether it's got two or three ( ″ / ‴ ) strokes beside it).* The filling-in method applies from p. 6 to p. 22.

2. *Points where you can take a breath* are shown by a comma. When taking a breath between two notes, it's the *first* of the two that's shortened. E.g: Bar 4, previous page, where the note is stopped with the tongue, followed by a breath, at '(four)'.

## Melody playing
### on the recorder

If you've played the accomps from the beginning, you've already got the notes you need for this tune:

OLD McDONALD p.24

Yours *Floppy*

## Counting method

Beneath the top 'line' of music you'll find a method for learning to count the beats for *the melody rhythm* (and which can also support the recorder's accomp line). This applies from p. 6 to p. 22. Count and clap the rhythm of each 'voice' or 'part'. Compare with 'Note values' on the pull-out page.
(See too 'General method' p. 3.).

# LONESOME VALLEY
American folk-song

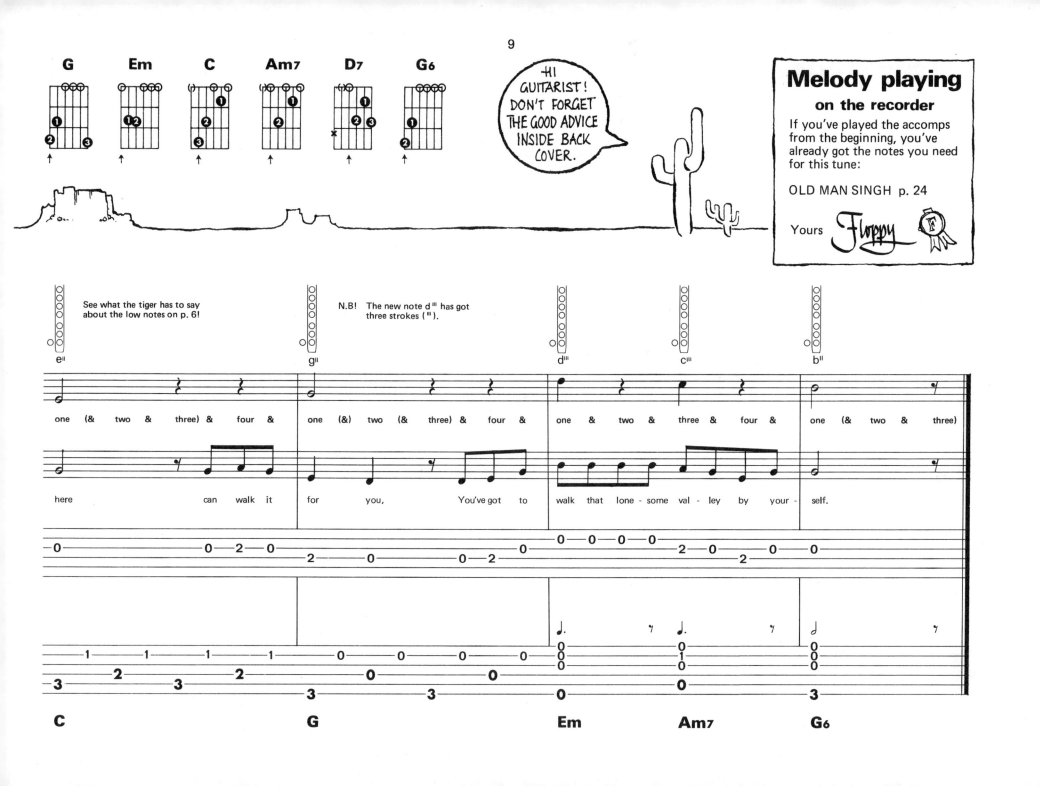

# BARGEE SONG

Russian folk tune

**Em**    **Am**

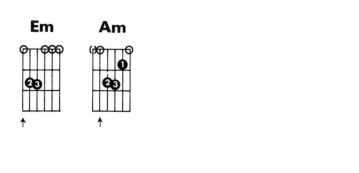

Interlude to be played if you like.

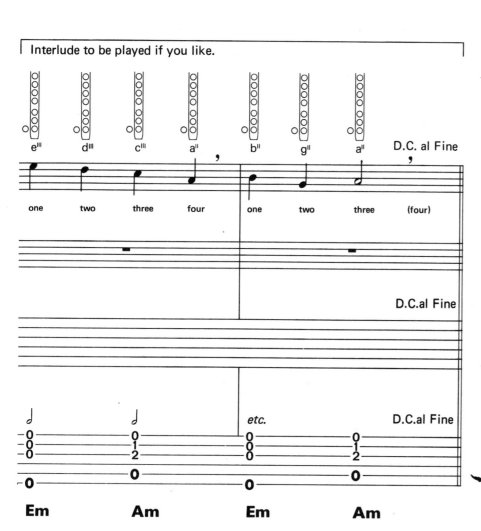

e'''    d'''    c'''    a''   ,    b''    g''    a''   ,    **D.C. al Fine**

one   two   three   four    one   two   three   (four)

D.C. al Fine

etc.     D.C. al Fine

**Em**     **Am**     **Em**     **Am**

## Guitar tips

The guitar chords to the right are constructed in such a manner as to facilitate a swift switching between G and $\frac{C}{g}$.

This, in turn, calls for *p-i-m-a-* playing with the right hand.

And, as always, do think of the advice inside the back cover.

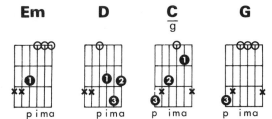

**Em**  **D**  **$\frac{C}{g}$**  **G**

p i m a   p i m a   p i m a   p i m a

## Melody playing
### on the recorder

If you've played the accomps from the beginning, you've already got the notes you need for this tune:

BARGEE SONG p. 25

Yours *Floppy*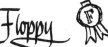

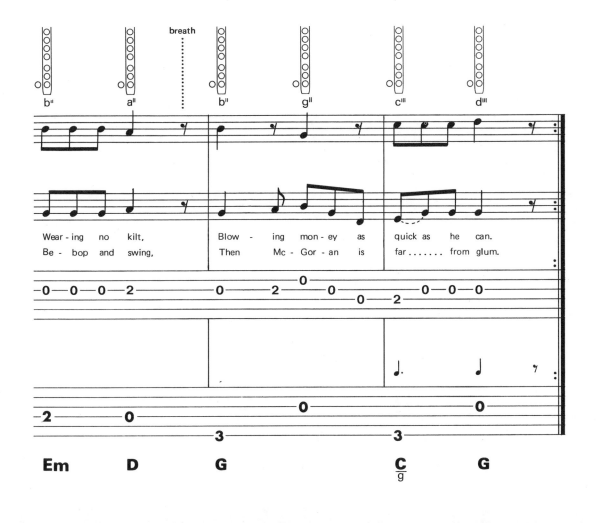

b" a" breath b" g" c"' d"'

Wear - ing no kilt,    Blow - ing mon - ey as    quick as he can.
Be - bop and swing,    Then Mc - Gor - an is    far....... from glum.

**Em**  **D**  **G**    **$\frac{C}{g}$**  **G**

SARDINES

# RAILROAD BILL

American folk-song

d" f#" g" a" b"

Fill in the right notes on the note-finders, using the Note-Finder System (see 'Recorder tips' p. 7).

Verse 3 : Kill me a chicken, send me a wing,
Think I'm working but I ain't doin' a thing,
Then it's ride, ride, ride.

Verse 4 : Railroad Bill, desperate and bad,
Take everything that your woman had,
Then it's ride, ride, ride.

Verse 5 : I'm going home, tell my wife
Railroad Bill tried to take my life,
Well, it's bad Railroad Bill.

Verse 6 : Railroad Bill, mighty bad man,
Killed McMillan with a gun in each hand,
Then it's ride, ride, ride.

Verse 7 : Railroad Bill, Railroad Bill,
He never worked and he never will,
He just ride, ride, ride.

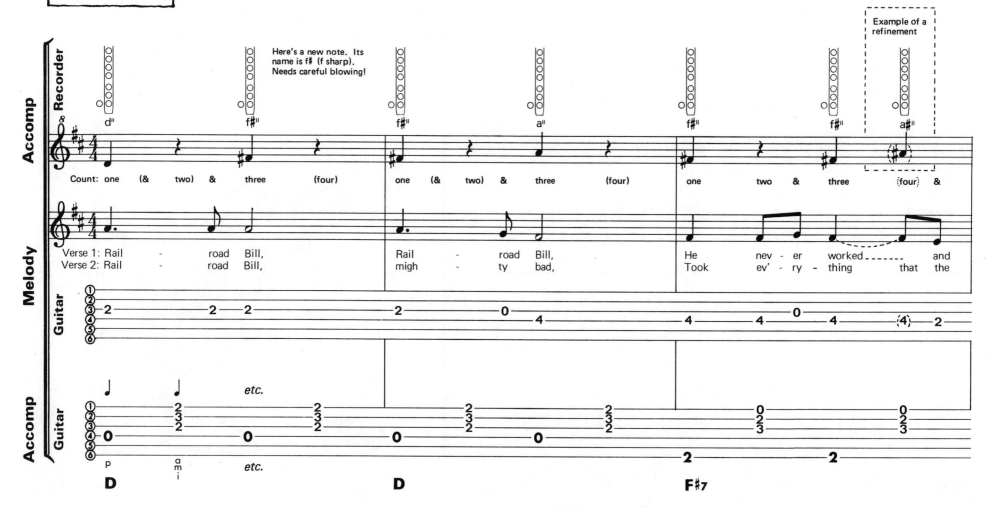

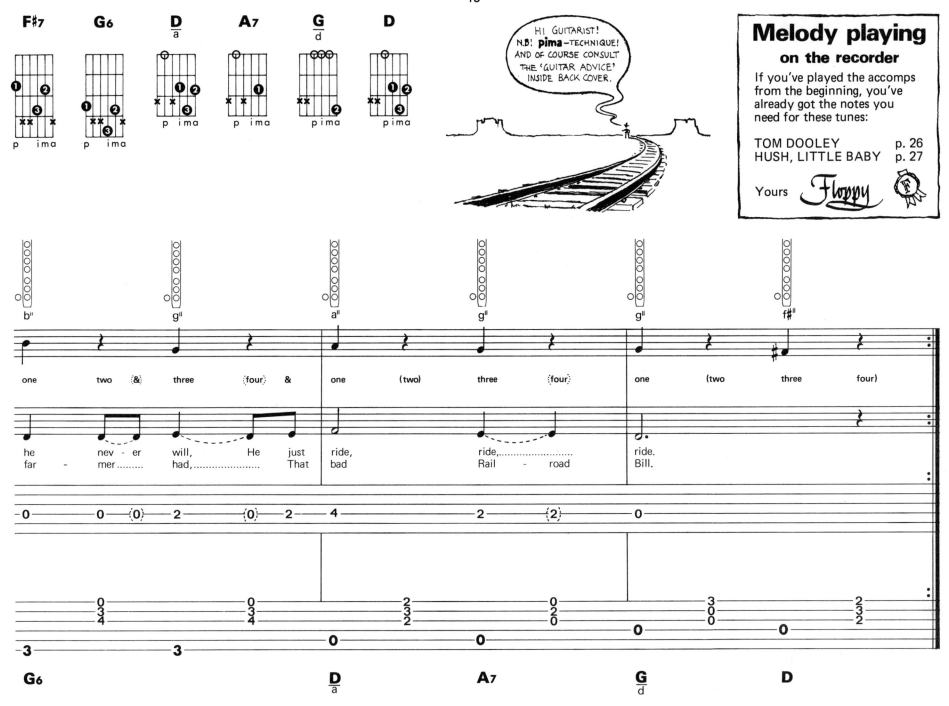

# SKYE BOAT SONG

Old Scottish sea-shanty
Words by Harold Boulton

d" e" f#" g" b" c" d"

Fill in the right notes on the note-finders, using the Note-Finder System (see 'Recorder tips' p.7).

Verse 2 : Though the waves leap, soft shall ye sleep,
Ocean's a royal bed.
Rocked in the deep, Flora will keep
Watch by your weary head. (Refr.)

Verse 3 : Many's the lad fought on that day,
Well the claymore could wield,
When the night came, silently lay,
Dead on Culloden's Field. (Refr.)

Verse 4 : Burned are our homes, exile and death
Scatter the loyal men;
Yet ere the sword cool in the sheath,
Charlie will come again. (Refr.)

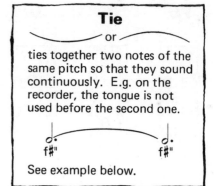

**Tie**

ties together two notes of the same pitch so that they sound continuously. E.g. on the recorder, the tongue is not used before the second one.

See example below.

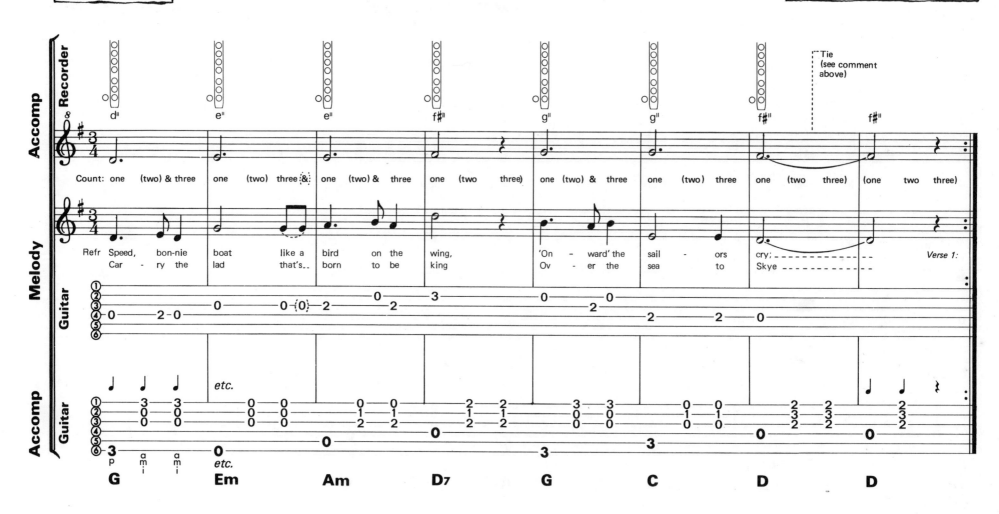

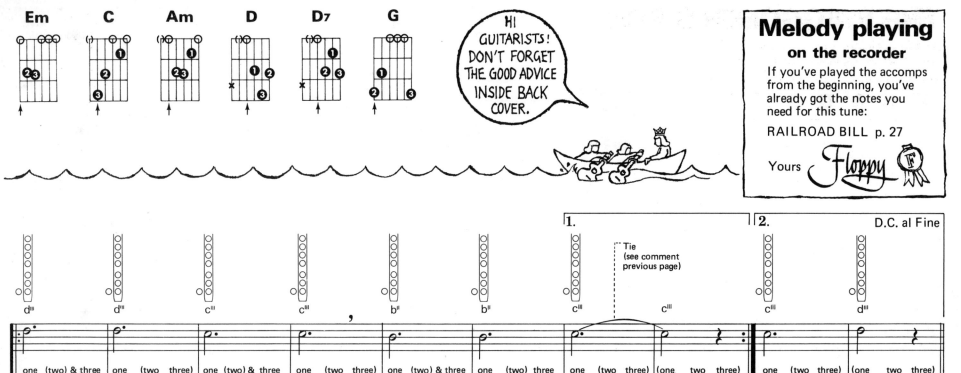

## Melody playing
### on the recorder
If you've played the accomps from the beginning, you've already got the notes you need for this tune:

RAILROAD BILL p. 27

Yours Floppy

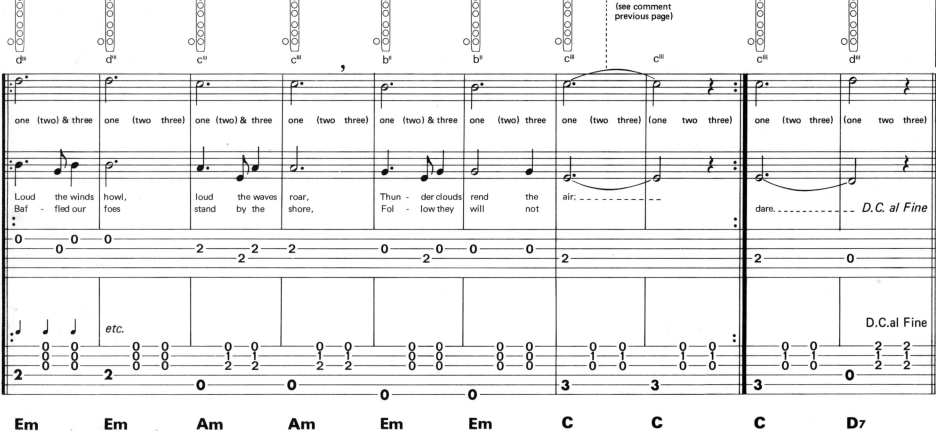

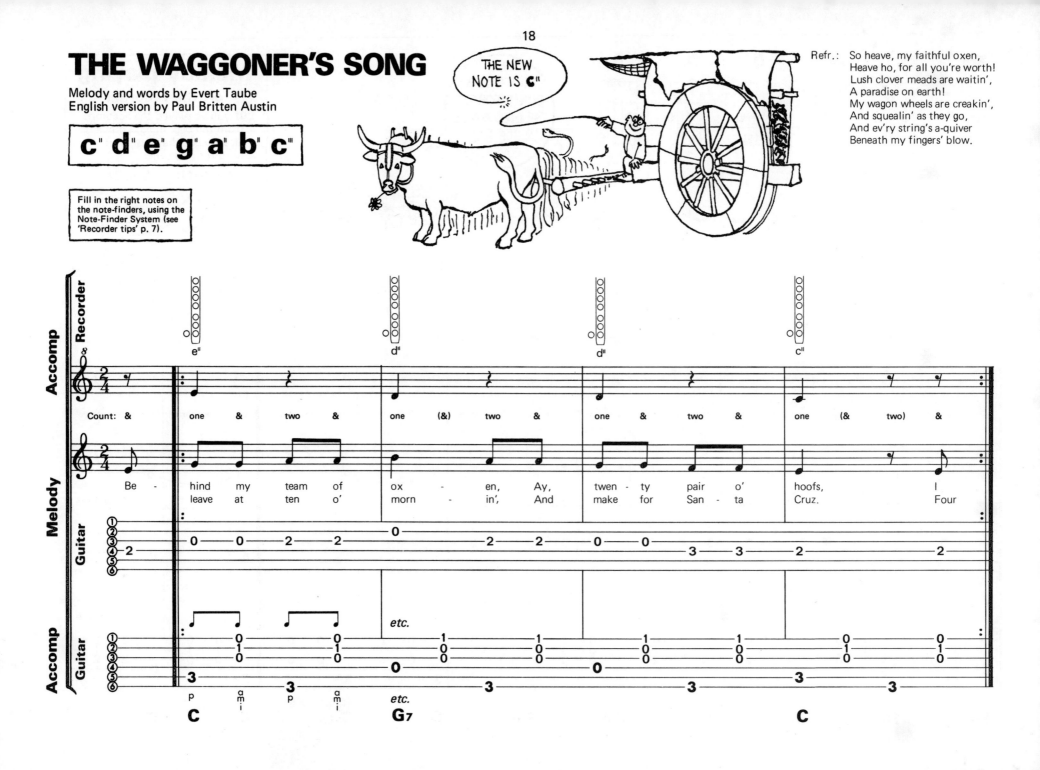

## Recorder tips

The new note's name is c'', in other words, time to stop all the holes. Oddly enough, this is done easiest if you just lay the finger-tips *lightly* on top of them.

Remember to blow gently, too, with a cautious 'DOOT' from your tongue!

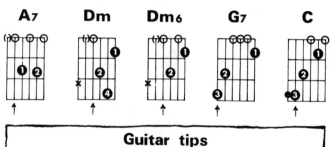

## Guitar tips

To make it easier you can play Dm6 instead of Dm. And of course you have read what is said under 'Guitar advice' inside the back cover.

## Melody playing
### on the recorder

If you've played the accomps from the beginning, you've already got the notes you need for these tunes:

McGORAN  p. 28
SKYE BOAT SONG  p. 28

Yours *Floppy*

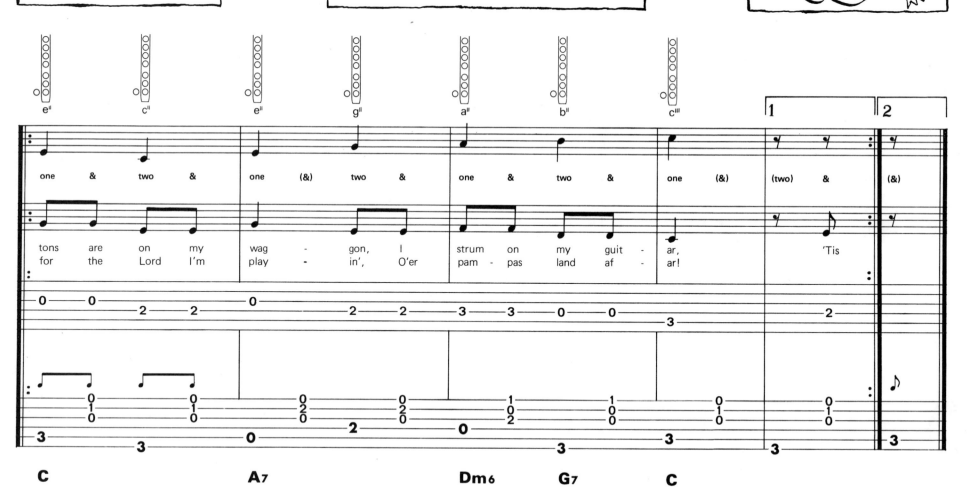

# NOBODY KNOWS

Negro Spiritual

**d″ f#″ e″ g″ a″ b″ c‴ d‴**

Fill in the right notes on the note-finders, using the Note-Finder System (see 'Recorder tips' p. 7).

### Melody rhythm

Compare the melody rhythm in the first bar with the same bar in 'Tom Dooley' p.2. In the same way you are of course free to vary the rhythm as is usual in this kind of melody.

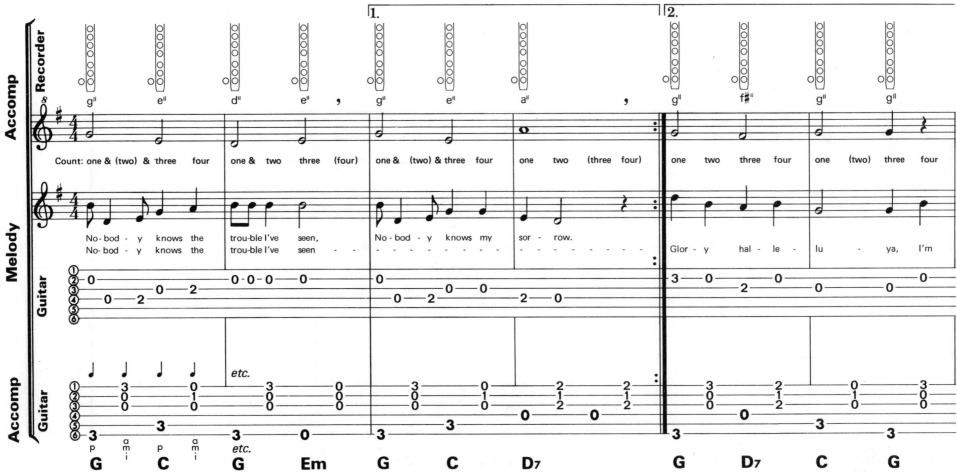

# MELODY PLAYING WITH CHORD SYMBOLS

**Melody rhythm**

You'll find the method for counting out the melody rhythms in the Accomp Section, pp. 2 - 22.
(See also page references beside each melody.)

**Guitar chords**

You'll find the diagrams for the guitar symbols on pp. 2 - 22.
(See also page references beside each melody.)

**Spain**

Chord diagrams: see p. 21

**To Paris**

Chord diagrams: see p. 21

# MELODY PLAYING WITH CHORD SYMBOLS

**Old McDonald**   Chord diagrams and melody rhythm pp. 4 & 5.

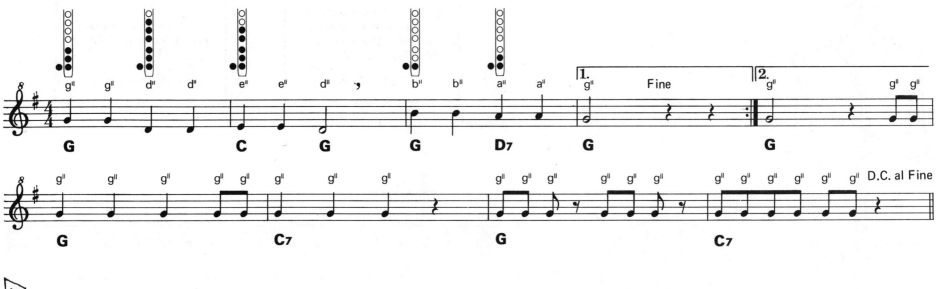

**Old Man Singh**   For the accompaniment and melody rhythm see pp. 6 & 7.

Lasse Sundberg

# MELODY PLAYING WITH CHORD SYMBOLS

**Lonesome Valley**    Chord diagrams and melody rhythm pp. 8 & 9.

**Bargee Song**    Chord diagrams and melody rhythm pp. 10 & 11.

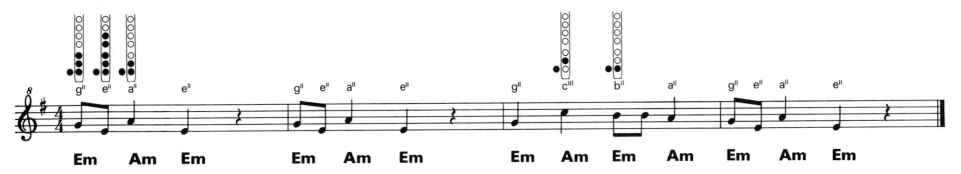

# MELODY PLAYING WITH CHORD SYMBOLS

**Tom Dooley** Chord diagrams and melody rhythm pp. 20 & 21.

# MELODY PLAYING WITH CHORD SYMBOLS

**Hush, Little Baby**  Chord diagrams and melody rhythm pp. 20 & 21.

**Railroad Bill**  Chord diagrams and melody rhythm pp. 14 & 15.

# MELODY PLAYING WITH CHORD SYMBOLS

**McGoran**  Chord diagrams and melody rhythm pp. 12 & 13.                         Ulf Goran Åhslund

FROM NOW ON YOU MUST FILL IN THE NAMES OF THE NOTES YOURSELF. SEE ALSO p. 7 'RECORDER TIPS'.

**Skye Boat Song**  Chord diagrams and melody rhythm pp. 16 & 17.

# MELODY PLAYING WITH CHORD SYMBOLS

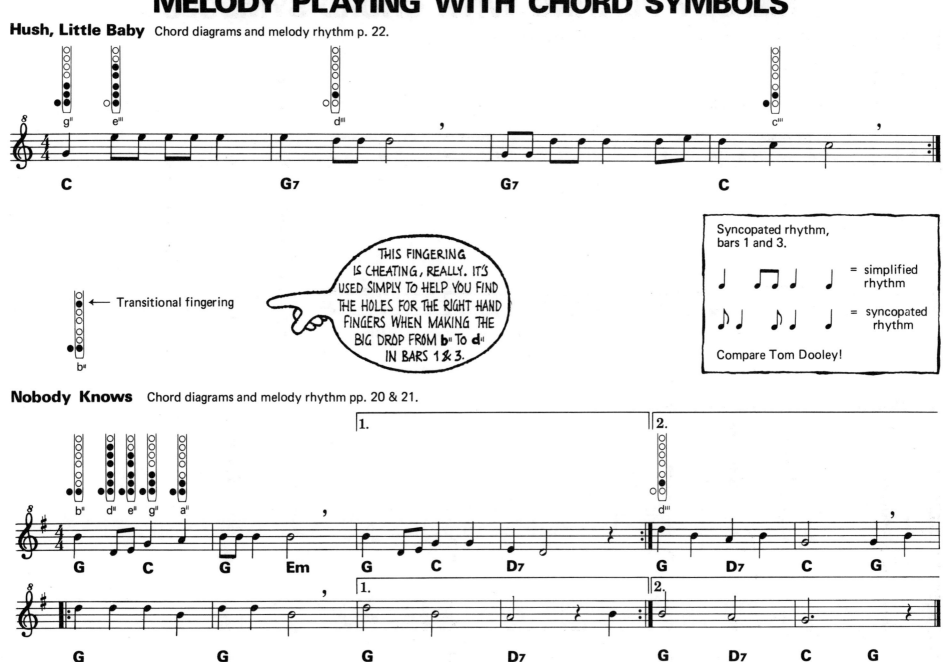

# MELODY PLAYING WITH CHORD SYMBOLS

**The Waggoner's Song**  Chord diagrams and melody rhythm pp. 18 & 19.

Evert Taube

**Tom Dooley**  Chord diagrams pp. 3 & 5:  melody rhythm p.2.

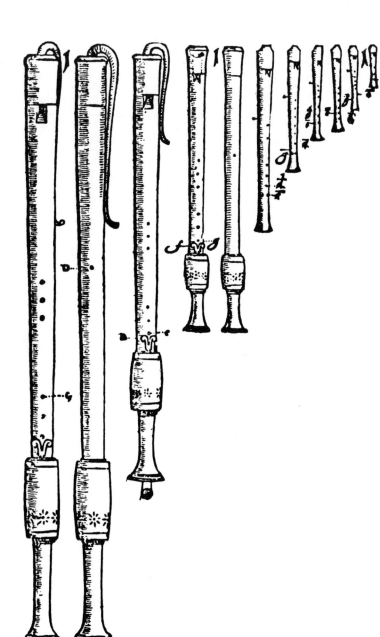

# ABOUT RECORDERS

The recorder is one of the oldest instruments known to us. Music has been written for it since the Middle Ages. By the Renaissance it had grown into such favour that recorders were being made in some ten different sizes and pitches. During the seventeenth century the 'F' treble recorder became the accepted solo instrument. But towards the end of the eighteenth century the recorder died out almost completely, being thrust aside by the transverse flute - what is today ordinarily called a flute. Now, after a new renaissance in recorder-playing, recorders are manufactured in many different sizes, but the four commonest pitches are the descant, treble, tenor, and bass. The lowest note of each is respectively c", f', c' and f.

## How to look after your recorder

A recorder is a sensitive musical instrument. If its tone is to stay clear and beautiful, then it demands a certain amount of regular care.

1. A new instrument has to be 'played in'. During the first week you shouldn't play it for more than at most ten minutes at a time, and then only two or three times a day. Gradually you can play it as much and as long as you like. Sometimes a new recorder will suddenly develop a very coarse tone. This is usually due to breath-condensation in the wind-channel. Lay one finger in the sound-hole (careful not to damage the lip of the instrument!), blow hard, and the coarseness will disappear.

2. The lip must always be carefully protected. The least damage can make the whole mouthpiece unusable.

3. Your recorder must be protected against excessive changes in temperature and from strong heat (direct sunlight, radiators, etc.). Otherwise it can split.

4. At least once a week, even more often to begin with, the joint must be smeared with vaseline or a special joint grease—otherwise it will tend to become tight.

5. About twice a year the tube, with the fingerholes, should be thoroughly lubricated with a special acid-free oil known as instrument oil, e.g. oil of almonds; that is, if it has not already been impregnated with paraffin wax during its manufacture. Soak a small (N.B!) rag in oil and attach it to a piece of string, then pull it through the tube a few times. But the mouthpiece must *never* be oiled!

6. Now and again it may be necessary to clean the wind-channel in the mouthpiece. But this should only be done by an expert, *not* by you.

7. Don't let other people play your recorder, except in very special circumstances. Your recorder soon gets used to just your own particular way of playing it, and easily suffers damage from other people's.

# DIAGRAM

*Diagram* = picture of strings and frets of neck.
*Black dots* = show where you must press down strings.
*Rings* = open strings belonging to the chord.
*Figures in diagram* = left-hand fingering.
*Letters under diagram* = right-hand fingering.

# LEFT HAND

The *left-hand fingers* are indicated by internationally accepted figures (see below).

# RIGHT HAND

The *fingers of the right hand* are indicated by the internationally accepted *p, i, m, a,* which are the first letters of each finger's Latin name.

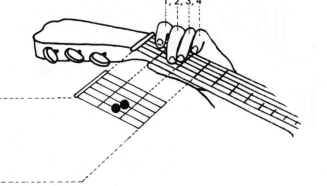

Chord ——▶ **E minor (Em)**

Open string ——▶

Left-hand fingering ——▶

Right-hand fingering ——▶ p

---

# VARIOUS BASS NOTES

**C**

↑ = Fundamental bass note (i.e. in the C-chord it's the note c.
✗ = The string must not sound in the chord.

**C/g**

Recommended alternative bass note that does appear in the chord symbol (indicated by a little letter - see chord symbol above)

**C**

········optional alternative bass notes········ that do not appear in the chord symbol.
● = small dot without fingering (for depressed string)
( ) = bracket (for open string)

**C**